Zoo Map

Dona Herweck Rice

Publishing Credits

Rachelle Cracchiolo, M.S.Ed., *Publisher*
Conni Medina, M.A.Ed., *Managing Editor*
Nika Fabienke, Ed.D., *Content Director*
Véronique Bos, *Creative Director*
Shaun N. Bernadou, *Art Director*
Seth Rogers, *Editor*
John Leach, *Assistant Editor*
Courtney Roberson, *Senior Graphic Designer*

Image Credits: All images from iStock and/or Shutterstock.

Teacher Created Materials
5301 Oceanus Drive
Huntington Beach, CA 92649-1030
www.tcmpub.com

ISBN 978-1-4938-9877-0

Printed in Malaysia
Thumbprints.42805

Find the 

monkey

in its  .

tree

Which way do

we go?

Find the 

bear

in its  .

cave

Which way do

we go?

Find the 

hippopotamus

in its  .

pond

Which way do

we go?

Find the 

bird

in its  .

nest

Which way do

we go?

Find the 

lion

in its  .

den

Which way do

we go?

High-Frequency Words

New Words

do	find
go	its
way	we
which	

Review Words

in	the